W9-ATY-445

EARTH, SPACE, AND BEYOND

WHAT DO WE KNOW ABOUT STARS AND GALAXIES?

John Farndon

Chicago, Illinois

an imprint of Capstone Global Library, LLC
Chicago, Illinois

Visit our website at
www.heinemannraintree.com

Edited by Andrew Farrow, Adam Miller and Adrian Vigliano
Designed by Marcus Bell
Original illustrations ©Capstone Global Library 2011
Illustrated by KJA-artists.com
Picture research by Hannah Taylor
Originated by Capstone Global Library Ltd.
Printed in China by South China Printing Company, Ltd.

15 14 13 12 11
10 9 8 7 6 5 4 3 2 1

Library of Congress Cataloging-in-Publication Data
Farndon, John.
What do we know about stars & galaxies? / John Farndon.
p. cm.—(Earth, space & beyond)
Includes bibliographical references and index.
ISBN 978-1-4109-4162-6 (hc)—ISBN 978-1-4109-4168-8 (pb) 1. Stars. 2. Galaxies. I. Title.
QB801.F37 2012
523.8—dc22 2010040167

Acknowledgments
The author and publishers are grateful to the following for permission to reproduce copyright material: Corbis pp. 4 (© Aflo Relax/ Sakuno Picture Agency), 9 (©Schlegelmilch), 13 (©Science Faction/ Peter Ginter), 21 (©NASA), 25 (©Robert Harding World Imagery), 34 (©NASA/ STScl), 35 (©epa/ Francis Specker), 42 (©Bettmann); NASA pp. 10 (JPL-Caltech), 12 (Hubble Heritage Team/ AURA), 14 (ESA/ Hubble Heritage Team), 16 (JPL-Caltech/STS cl), 17, 18 (ESA/JHU/R.Sankr it & W.Blair), 20 (CXC/Caltech/S.K ulkarni et al./ STScl/ UIUC/Y.H. Chu & R.Williams et al/ JPL-Caltech/R.G ehrz et al.), 23, 26 (The Hubble Heritage Team (STScl/AURA), 29 (JPL-Caltech/R. Hurt [SSC]), 30 (ESA/ M. J Jee [John Hopkins University]), 31, 32, 33 (CXC/MIT/F.K.Bag anoff et al.), 37, 43, 38 (CXC/M.Weiss), 41 (ESA/JPL-Caltech/ STSc I/D. Elmegreen [Vassar]); Science Photo Library pp. 7 (©NASA/ESA/STSCI/R.Kennicut, U.Arizona), 8 (©Larry Landolfi), 19 (©Mark Garlick), 22 (©Jason T Ware), 24 (©John Chumack), 27 (©ASA/ESA/B Whitmore/STSCI-AURA), 28 (©Science Source), 40 (©Mark Garlick), 39 (©NASA/ESA/STSCI/Hubble Heritage Team).
Cover photograph of spiral galaxy Messier 101 (M101) reproduced with permission of NASA (CXC/JHU/K.Kuntz et al).

We would like to thank Professor George W. Fraser for his invaluable help in the preparation of this book.

EARTH, SPACE, AND BEYOND

WHAT DO WE KNOW ABOUT STARS AND GALAXIES?

Contents

Some words are shown in bold, **like this**. You can find out what they mean by looking in the glossary. You can also look out for them in the "Word Station" box at the bottom of each page.

What Can We See in the Sky?

If you look into the sky on a clear night, you'll see the Moon and thousands of twinkling pinpoints of light spread across the darkness. All these twinkling lights are stars. Stars twinkle because dust and gas surrounding Earth get in the way of your view, and also because the atmosphere itself fluctuates. You may see one or two pinpoints of light that don't twinkle. These are planets, our neighbor worlds that circle the Sun along with the Earth. Although much smaller than stars, planets are so much closer to Earth that they form pinpoints of light too big to be blocked out by dust. This is why you won't see planets twinkling.

How many stars are there?

You can see just a few thousand stars with your eyes alone. With an amateur astronomical telescope you can see a few million. But the Universe is beyond enormous. All of the matter and space in existence make up the Universe. Every star in the Universe is part of a giant star city called a **galaxy**. Astronomers have found that there are about 100 billion stars in each galaxy. They have also found about 100 billion galaxies in the Universe. That means there must be at least 100 billion times 100 billion stars in the Universe!

Fixed stars

There is another difference between stars and planets. If you watch the sky for long enough, it looks as if all the stars are moving slowly from east to west. In fact, it is not the stars moving but the Earth. If the Earth stopped spinning, you'd see stars stay perfectly still, in a fixed pattern. It takes 24 hours for the Earth to turn around. So the star pattern returns to the same place every 24 hours.

Unlike the fixed stars far away, however, the planets actually move through the sky as they circle the Sun close by. That's how **astronomers** in ancient times knew planets were different from stars, even though both were just pinpoints of light. Today, we can actually see that planets are globes like the Moon.

The stars are much too far away to ever visit, and even the most powerful telescopes only show them as pinpoints of light. However, telescopes have revealed the secrets of the stars in other ways.

In a photo that uses a long exposure, such as this one, the movement of the stars through the sky as the Earth turns means that the stars appear as long curving streaks.

How far are the stars?

One of the biggest challenges astronomers face is measuring distances to stars far away across space. They can calculate how far a nearby star is by the **parallax** method. As the Earth circles the Sun, the star shifts sideways slightly compared to stars further away, as the angle we see it from changes. The nearer the star is to Earth, the more it shifts.

With distant stars, this side-shift is too small to measure. The secret is to compare brightness. A star's relative **magnitude** is how bright it looks compared to others. But a dim star might be shining feebly — that is, it has a low absolute magnitude. Or it may just be far away. We can tell which by its color. The whiter a star is, the hotter and brighter it glows; the redder it is, the cooler and dimmer it is. A dim white star is a star far away. A bright red star is a star nearby. Astronomers have made a graph showing the relationship between color and brightness for most ordinary or Main Sequence stars. Calculating distances using this graph is known as Main Sequence fitting.

This diagram shows a basic example of the parallax method. By measuring the angles of a star's light from two points in Earth's orbit, an astronomer can calculate how far away the star is. The further away the star is, the smaller the parallax angle.

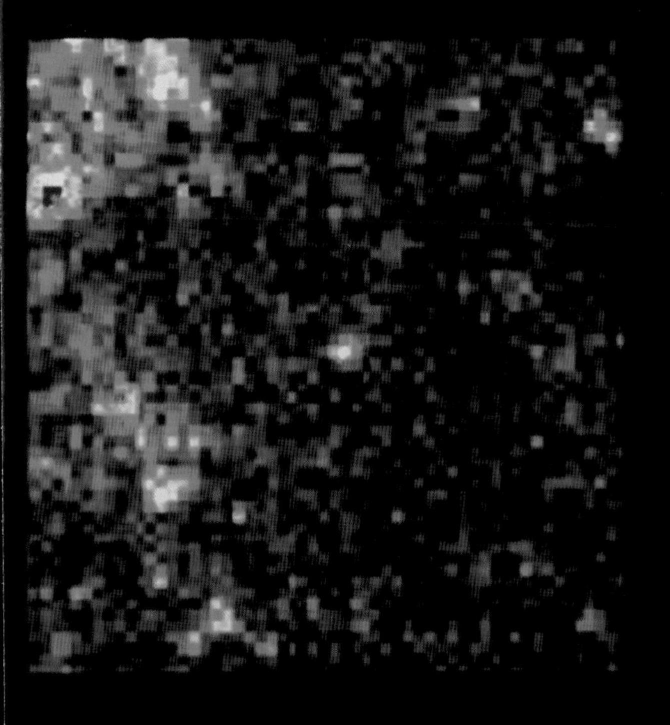

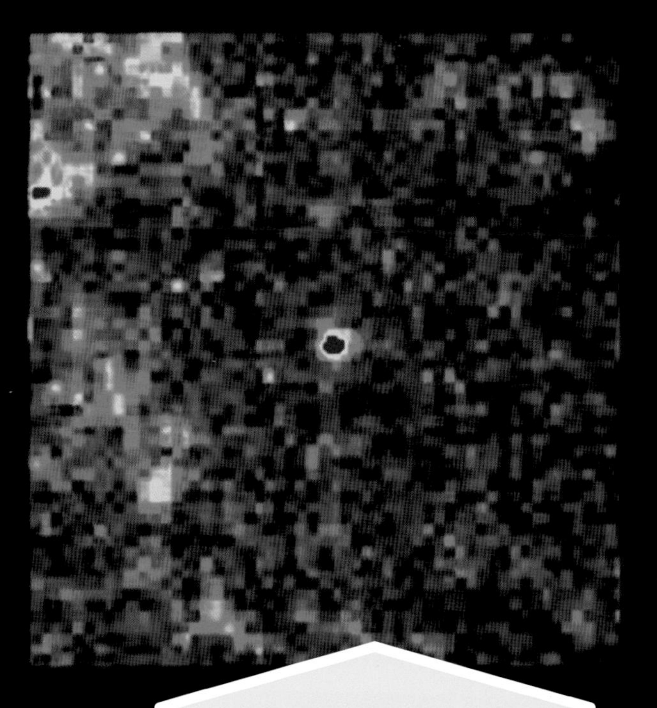

These two images of Cepheid **variable star** M100 show how it varies in brightness. The speed of variation indicates its true or absolute brightness, enabling astronomers to calculate that it is 56 million light-years away.

Standard candles

Beyond about 300,000 light-years, stars are too blurry to use even Main Sequence fitting. So to measure the distance to other galaxies, astronomers look for standard candles — stars they know the brightness of for certain. For nearby galaxies, the best standard candles are stars that appear to flash on and off at regular intervals, known as **Cepheid** variables. The brighter a Cepheid is, the longer the time between flashes. So astronomers can tell how bright a Cepheid should really look, and how far away it is, by timing its flash.

With very distant galaxies, astronomers can't see single stars so must use other methods. In the Tully-Fisher technique, for instance, they work out how bright a galaxy should be from the speed it is spinning. They then compare this to how bright it actually looks.

The brightest star

Sirius is the brightest looking star in the sky by relative magnitude. But, a star called the Pistol star is actually 400,000 times brighter — 10 million times brighter than our Sun in absolute magnitude. But the Pistol star is so far away it actually looks quite dim.

WORD STATION

magnitude scale for comparing the brightness of stars

Measuring by angle

Astronomers also use another measurement for distances to the stars, the parsec. The parsec came from the angle a star shifts when measuring its distance with the parallax method, but it is now used whatever way the distance is measured. A parsec is about 31 trillion kilometers (19 trillion miles). Proxima Centauri, the nearest star, is about 1.3 parsecs away.

Written in the stars?

Distances to stars and galaxies are so vast it's not practical to measure them in miles or kilometers. Even the nearest star, Proxima Centauri, is over 40 trillion kilometers (25 trillion miles) away. The furthest galaxies are 12 billion trillion kilometers (7 billion trillion miles) away. Instead, astronomers measure distances in light-years.

We see stars by the light they beam out to us, but the distance is so large that their light takes years to reach us. Light always travels at the same speed, 299,792 kilometers per second (186,282 miles per second). So counting the years the light takes to reach us is a good way to measure distance. A light-year is the distance light travels in a year, which is 9,460 billion kilometers (5,878 billion miles), or 0.3 **parsecs** (read more about parsecs at left). Light takes 4.2 years to reach us from the star Proxima Centauri, so it is said to be 4.2 light-years away.

The lines mark out the stars in the constellation known as Cygnus the Swan. The star on the left is Deneb, which must be huge since it looks very bright, even though it is 1,500 light-years away.

When a racing car zooms past, the whine of its engine falls in pitch as it races away from you. This is because the sound waves are stretched out in what is called the Doppler effect. The redshift of receding stars is a similar effect.

A receding star, with waves stretched to red

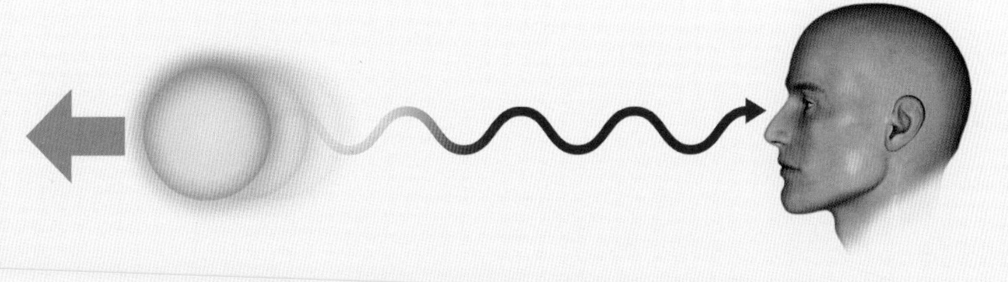

An approaching star, with the waves compressed (squashed) to blue

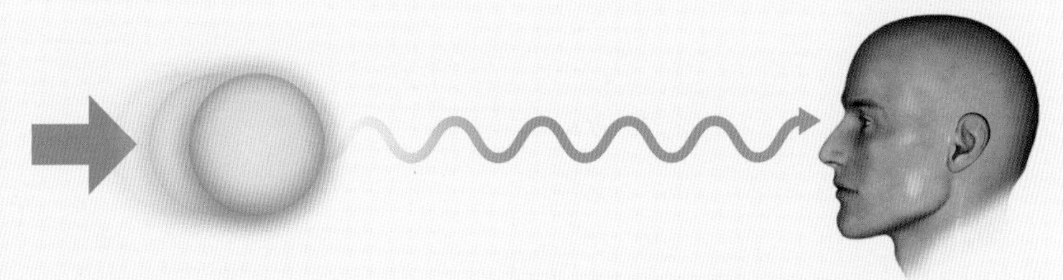

What's interesting is that if light takes 4.2 years to reach us, we are seeing the star not as it is now, but as it was 4.2 years ago. The bright star Deneb, for instance, is about 1,500 light-years away. So we are seeing it as it was at the beginning of the 6th century CE, not long after the fall of the Roman Empire. With the most powerful telescopes we can see galaxies 13 billion light-years away, which means we are seeing them not long after the Universe was born.

Astronomer's trick: How to spot a star on the run

If a star is rushing away from us, each light wave from it gets sent from a little further away, so appears to get stretched out. As light waves stretch, they get redder. This is called a Doppler shift or **redshift**. If the star is moving toward us, the light waves get compressed and bluer (a **blueshift**). So you can tell how fast a star is moving toward us from the amount it is red or blueshifted. Astronomers know galaxies are spinning because stars are blueshifted on one side and redshifted on the other.

WORD STATION

parsec measure of distance in space equal to 3.3 light-years

Just what can we see?

Over the last century, astronomers have created powerful telescopes to probe far into the Universe. A hundred years ago, the most distant object understood by astronomers was only 15,000 light-years away. Now astronomers can study galaxies 13 billion light-years away — almost a million times further. As they have seen further, windows have opened on billions of stars and galaxies.

As well as larger and better telescopes on the ground, astronomers also have telescopes on satellites in space, such as the Hubble Space Telescope. These give a view of space that is not clouded or distorted by the Earth's atmosphere. Sensitive digital image devices can also record light from sources much too faint for our eyes to see. Computers can process and enhance images to reveal hidden details.

The Pleiades or Seven Sisters is one of the brightest constellations. It is also one of the few that is a physical cluster of stars rather than just a collection of stars in roughly the same direction.

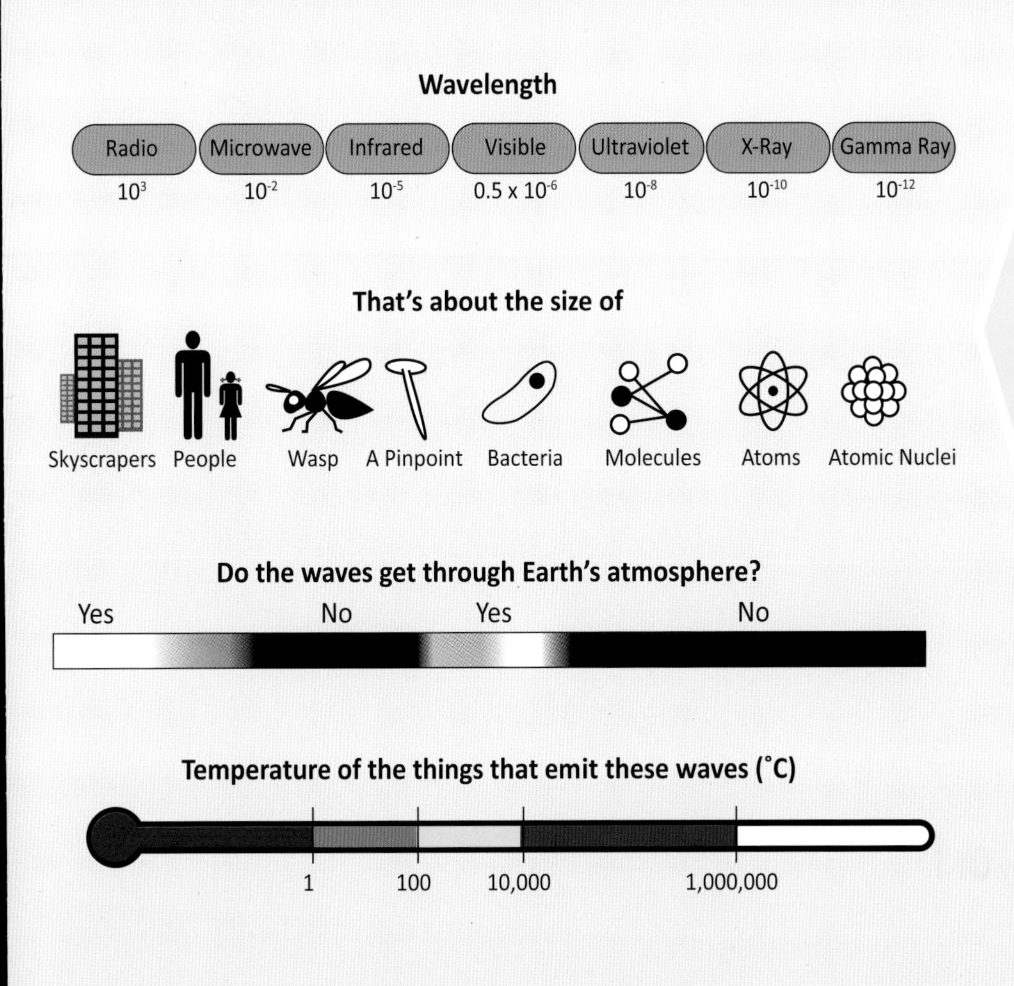

Different wavelengths in the electromagnetic spectrum range from huge to incredibly small. Use this diagram to compare facts about wavelengths.

Most telescopes use the visible light our eyes can see. But special telescopes can pick a range of other, invisible radiation, sent out by stars and galaxies, including radio waves, **infrared**, **X-rays**, and **gamma rays**. Antennae on the ground such as satellite TV dishes, for instance, can detect radio waves from space. These allow astronomers to see through the cloud of dust right to the heart of the **Milky Way** Galaxy and discover what's going on there. X-rays and gamma rays are blocked off by the atmosphere but they can be detected using space telescopes. X-ray telescopes have played an important part in discovering **black holes** and small dense stars called **neutron stars**.

Some telescopes can pick up the sub-millimeter waves and microwaves between radio waves and visible light. Sub-millimeter telescopes can see through dust clouds to reveal what's going on as stars form. Microwave telescopes can see the very furthest reaches of space to show the pattern of radiation soon after the Universe began.

The electromagnetic spectrum

Almost everything we know about the stars and galaxies we know because the stars and galaxies beam out radiation. The full range of radiation, including visible light, is known as the **electromagnetic spectrum**. It travels in waves, and the nature of each kind of electromagnetic radiation depends on its wavelength. Radio waves, microwaves and infrared are waves too long for the eye to see. **Ultraviolet**, X-rays, and gamma rays are too short for the eye to see. In between comes the visible light we can see.

WORD STATION

electromagnetic spectrum entire range of electromagnetic radiation, including light that is visible to humans

What Are Stars?

Stars are huge, fiery balls of gas. Astronomers know they are made mostly of hydrogen and helium gas from the color of the light coming from them. When stars burn, every **element** in the star glows with its own unique spectrum, or range of colors, which identifies it like a name tag. So, scientists can discover what stars are made of by analyzing the colors in starlight, a technique known as spectroscopy.

Spectroscopy also shows that most stars also contain traces of iron and other heavier elements as well hydrogen and helium. These stars, known as Population I stars, are the Universe's second generation of stars. There are rarer, older stars, known as Population II stars, that contain only hydrogen and helium. These were the first elements to be made when the Universe was young. The heavier elements like iron were created as hydrogen and helium atoms combined under pressure inside Population I stars.

In this image of the Egg Nebula, some of the layers, or "shells" of dust that block the central star from view can be seen. Outer layers of dust reflect light from the star, creating the effect seen here.

To detect neutrinos streaming from the Sun, scientists are preparing 5,000 detector balls and encasing them over 1,000 meters deep in the Arctic ice in a huge array known as the Ice Cube.

CHASING NEUTRINOS

To prove the nuclear fusion theory of stars, scientists needed to detect particles called neutrinos coming from the Sun. Yet neutrinos are so tiny they are hard to spot among many other larger particles at the Earth's surface. But their small size means they can fly clean through solid rock that stops other particles. So scientists built their detectors far underground. The problem was then how to stop the neutrinos. In the 1960s, Scientists John Bahcall and Ray Davis succeeded in detecting a few neutrinos in vast tanks of a chlorine cleaning fluid. In 1998 and 2001, Japanese scientists found evidence that neutrinos can change their type as they travel. This helped explain why scientists find so few neutrinos and led to new theories about how neutrinos work.

Why do stars shine?

Stars shine because they are heated to enormous temperatures by nuclear reactions, like giant nuclear bombs. Deep inside, hydrogen atoms are squeezed together so hard by the star's **gravity** that they fuse to form helium atoms, releasing energy. This is called nuclear **fusion**. So much energy is released that the star's heart reaches millions of degrees, and the surface glows white hot.

We know stars are powered by nuclear fusion partly from their size and their brightness. For their size, stars emit just the energy (shown by their brightness) that scientists would expect if they are powered this way. Extra proof came when scientists detected incredibly tiny particles called **neutrinos** coming from the Sun. Neutrinos are a by-product of the nuclear fusion of hydrogen.

WORD STATION

fusion release of energy caused by atoms of certain elements joining together under the pressure of gravity

What are giants and dwarfs?

- Red giant stars are massive old stars up to 50 times as big as the Sun, while supergiants are much bigger still. The star VY Canis Majoris is over 2,000 times as big as the Sun, and burns almost half a million times brighter.
- White dwarfs are still warm and shining remnants of old stars about the size of the Earth.
- Black dwarfs have stopped shining.
- **Brown dwarfs** were never big enough to even start shining in the first place.

What can you tell about a star?

Everything about a star – how hot it is, what color it is, and how long it will live – depends on its **mass**.

Stars are so far away it's not always possible to tell how big, or even how bright, they are. But you can tell a lot from their overall color. The colors are usually too pale to see with your own eyes, but they show up clearly on astronomers' recording equipment. The color varies with a star's temperature, from dull red and orange for cool stars, to bright white and blue for the hottest. The astronomers Ejnar Hertzsprung and Henry Russell created an important graph showing how bright and hot a star would be for every color.

VY Canis Majoris is a **red supergiant** star that is also classified as a hypergiant because of its very high emission of radiation. It is one of the brightest stars in the sky, and is nearing the end of its life.

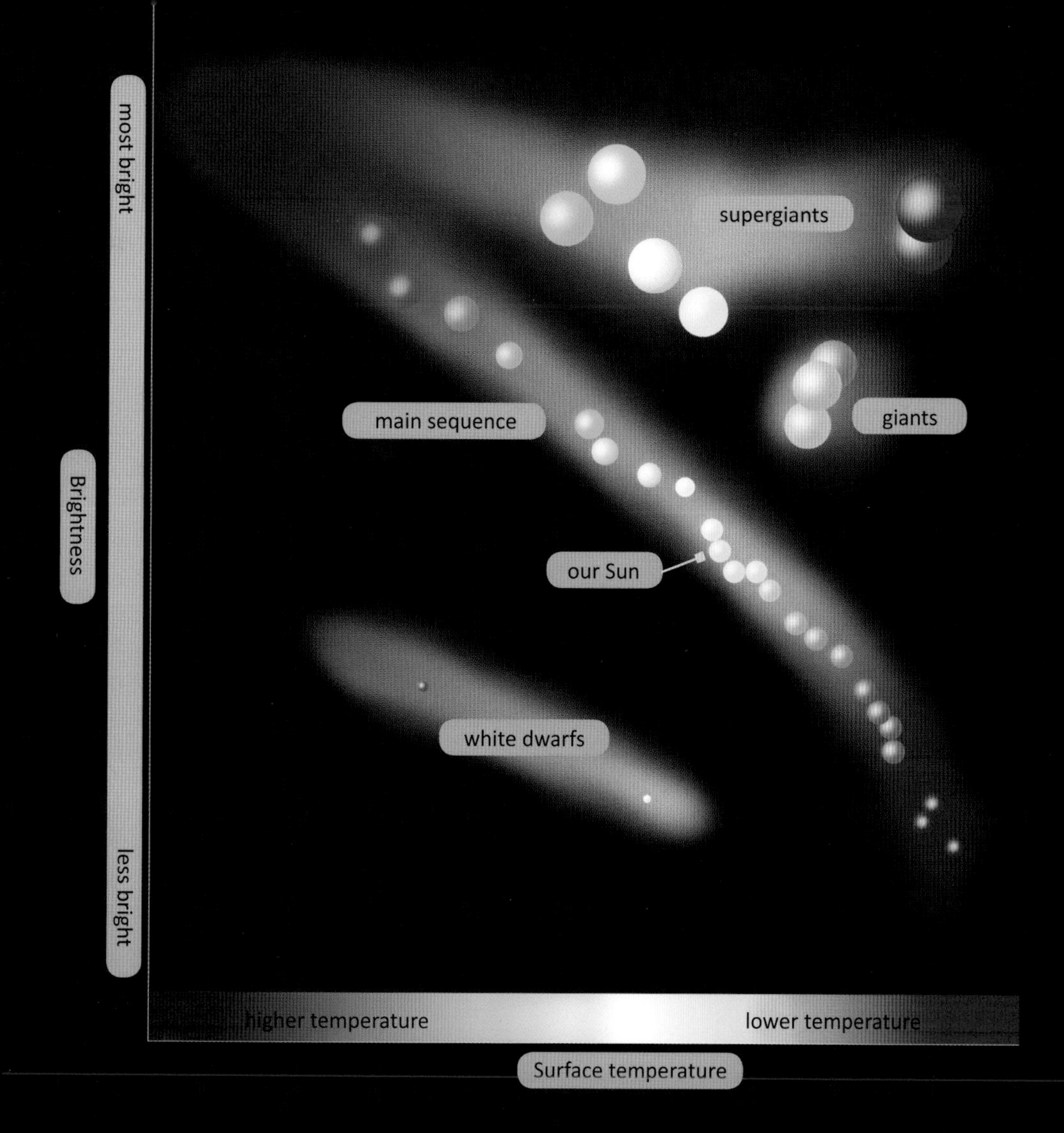

Which stars are red?

Using the diagram above, you can tell a star's size from its color. You can also use color to determine the star's temperature and true brightness. Stars that shine blue or white are hot, bright, and big. Stars that shine orange and red are cool, dim, and small. Stars that shine yellow or yellow-red are in the middle of the scale.

There is a diagonal band in the middle of the graph, known as the Main Sequence. All stars in the prime of life (when they are producing energy by fusing hydrogen atoms) fall into this band. Only stars near the end of their life, such as **white dwarfs**, red giants, and **supergiants** fall outside the graph.

Astronomers break Main Sequence stars into seven main groups according to their color: O, B, A, F, G, K, and M. The biggest, hottest stars are group O, the smallest and coolest are group M.

WORD STATION

mass measure of the amount of matter contained in an object

A Star Is Born

Stars are dying and being born all over the Universe. By looking at many stars, each at a different stage of its life, astronomers have pieced together the picture of the entire life of the stars.

Stars start life in big clouds of dust and gas called **nebulae**. Every now and then, pressure variations within the clouds create clumps. The clumps are then pulled together under their own gravity into dark blobs called **dark nebulae**. Each blob contains a family of baby stars or **protostars**. As gravity goes on squeezing, the protostars begin to get hot. Protostars that are quite small never get very hot and soon fizzle out as brown dwarfs. But if the core of a protostar reaches 10 million degrees C (18 million degrees F), hydrogen atoms start to fuse together in nuclear reactions, making the star glow brightly.

First discovered by Galileo in 1617, the Trapezium cluster is a dense cluster of stars only just visible through a cloud of dust and gas. It is likely that stars are being born here in this cloud.

In October 2010, astronomers discovered the most Earthlike planet yet, Gliese 581g (named because it is one of eight planets found circling the star Gliese 581). Gliese 581g is larger than the Earth, but may have water.

"Sometimes I think we're alone in the Universe, and sometimes I think we're not. In either case, the prospect is staggering!"

Famous writer Arthur C. Clarke, contemplating the possible existence of alien life

In medium-sized stars like our Sun, the heat of hydrogen burning pushes gas outward as hard as gravity pulls it in, creating a balance. So the star settles down to a long adult life, burning steadily for perhaps ten billion years until all its hydrogen fuel is burned up. Since it formed 4-5 billion years ago, the Sun is about half way through its life. Smaller stars burn dimly and redder, and may live for more than 200 billion years. The biggest, most massive stars live fast and die young, burning brightly, hot and blue, for just ten million years.

Earthlike planets around other stars

Our Sun is not the only star to have planets circling it. By June 2010, astronomers had detected 464 planets circling other stars. In fact, planets have been detected around 12 percent of all stars looked at. Most of these extrasolar planets cannot actually be seen, but are detected by the way the effect of their gravity makes their parent star wobble slightly.

The planets found are mostly big planets made of gas, not small solid planets like Earth. They orbit either too far from, or too close to, their star for water to exist as a liquid, and so be hospitable to life. But in April 2010, a planet called Gliese 581g, was shown to be just the right distance from the red dwarf star Gliese 581. If there are many more planets like these in the Universe, there is a high chance Earth may not be the only place where life has begun.

WORD STATION

nebula (nebulae) a cloud of interstellar gas and dust where the beginnings of stars are formed

This picture shows the remnants of the most recent supernova explosion in the Milky Way, observed by Johannes Kepler in 1604 and so known as Kepler's Supernova. It was made by combining shots from three separate telescopes.

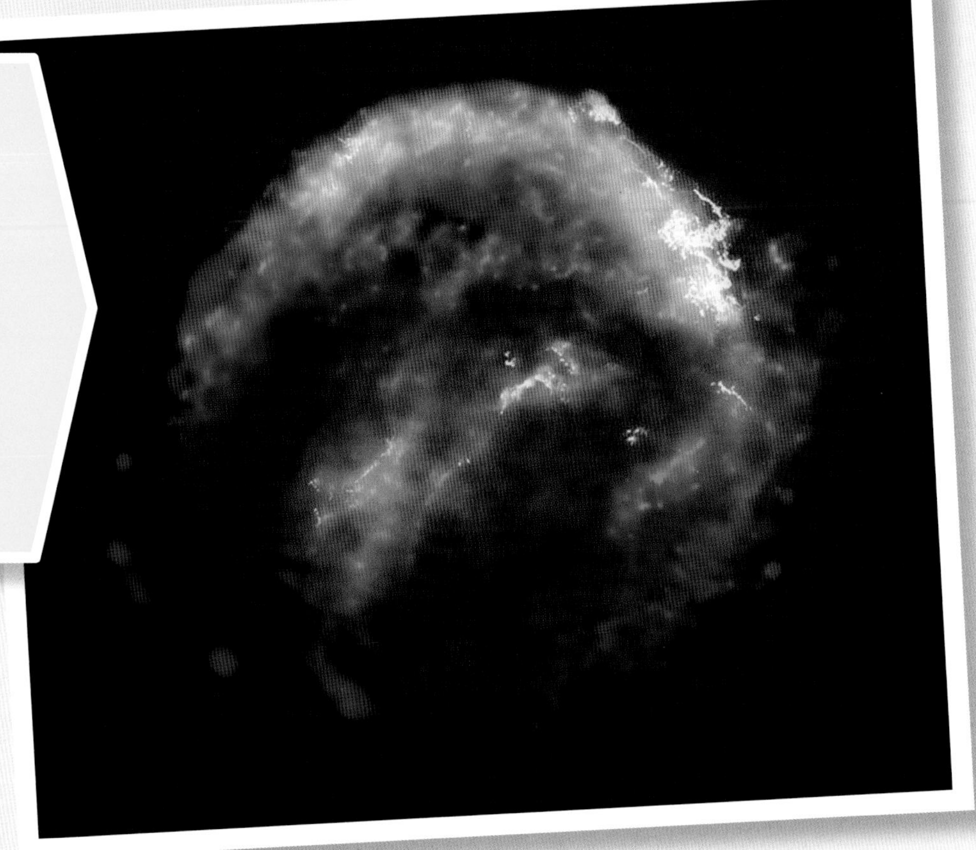

"White dwarfs, the slowly cooling remains of stars that have completed their life cycles, often seem to be the zombies of the night sky, devouring anything that happens to stray within their grasp."

www.blog.professorastronomy.com, June 25, 2010

How do stars die?

After about ten billion years, all the hydrogen fuel in a star's core is burned out. With no heat from the core to counteract gravity, the star begins to collapse. But hydrogen starts to fuse and burn in a thin layer surrounding the core. The heat puffs the star's outer layers of gas up enormously, so that the star expands to many times its original size. Small and medium-sized stars like our Sun expand to become cool red giants. Larger stars become red supergiants, such as Betelgeuse in the constellation Orion, which is 500 times as big as the Sun. The outer layers of red giants go on swelling more and more until they blow off completely, leaving behind the star's hot, dead core. This naked core is a white dwarf star. It lights up the surrounding cloud of blown away gas to create what is known as a glowing nebula. Over about 10,000 years the gas spreads out until it is too thin to see. The white dwarf cools and fades too, leaving just a cold black dwarf.

How do huge stars die?

Huge stars die far more dramatically. When a huge star's hydrogen fuel burns out, it begins to collapse so powerfully that the helium left in its core fuses to make carbon and oxygen. It gets very hot and swells to become a red supergiant and the carbon and oxygen fuse to make sodium, magnesium, silicon, and finally iron. Once iron forms, the star's core collapses so suddenly that it rebounds some way again — and smashes into the inward rushing outer layers. The smash blasts the remnants of the outer layer far out into space in a massive **supernova** explosion. The explosion is visible only for a week, but shines as bright as a galaxy of 100 billion stars.

The death star

In January 2010, astronomers announced that a huge white dwarf in the two-star system T Pyxidis was much closer than they'd thought, just 3,260 light-years away. White dwarfs can explode as supernovae. Newspapers called the star "The Death Star" and the British paper the *Sun* said, "A star primed to explode in a blast that could wipe out the Earth was revealed by astronomers yesterday." Of course, we are too far away to be affected by the blast in the way that newspapers like the *Sun* implied. But there is a slight danger, which may come from streams of gamma rays that could damage the ozone layer and expose Earth to radiation. But, the damage probably won't happen for another ten million years!

Gamma ray bursts (GRBs) are the brightest events in the Universe, releasing as much energy in 10 seconds as the Sun does in 10 billion years. This impression of a GRB was done by the astronomer Mark Garlick.

Alternative stars?

Stars don't all burn steadily like our Sun. Many tend to flare up and down, some at regular intervals, some irregularly. Stars like these are known as variable stars. Some, called Cepheids, are big stars that throb with energy, pulsating from every few days to every few months. They are used for measuring distances in space. RR Lyrae variables are tired, old yellow supergiant stars, so feeble they can only flicker. Some stars like Mira A change regularly over months or even years. Others vary with no obvious pattern.

In 1967, astronomer Jocelyn Bell picked up amazingly intense, regular radio pulses. For a while, astronomers joked they might even be signals from aliens and called them LGMs, short for "little green men." Soon, other LGMs were detected and it was realized that neutron stars might beam out these pulses as they spin, like a lighthouse, and so LGMs were renamed **pulsars**.

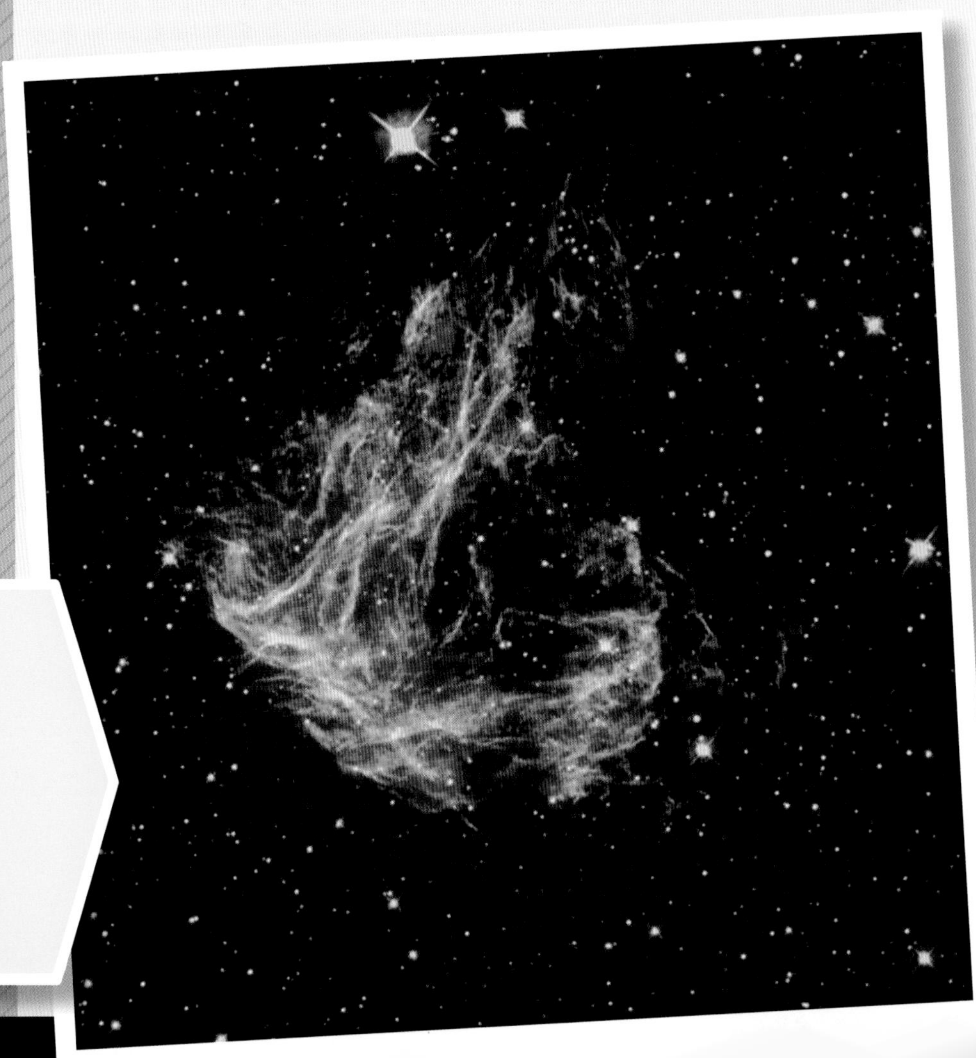

When the Chandra X-ray observatory took pictures of the N49 supernova it revealed what has been described as a cosmic bullet shooting away. This "bullet" is probably a neutron star.

This artist's impression shows the Cassini space probe arriving at Saturn's rings. It beamed back valuable data about the rings, as well as fuzzy signals which some have insisted are alien broadcasts, but are likely to be just natural emissions.

Neutron stars are all that's left of supergiant stars after a supernova. The star collapses so powerfully under its own gravity that all its matter is squeezed into a tiny glowing ball barely 20 kilometers (12 miles) across. These balls are made entirely of **neutrons**. Normal matter is made of three main kinds of particles: protons, electrons, and neutrons. In a neutron star the atoms are squeezed so tightly together that protons and electrons combine to form neutrons. Since so much matter is squashed into such a small space, a neutron star is incredibly dense. A teaspoonful of neutron star would weigh about 10 billion tons!

Only a few neutron stars are pulsars. After spinning for several million years pulsars are drained of their energy and become normal neutron stars.

Aliens speaking?

In June 2007, someone had the idea of using a computer to strengthen some radio signals from NASA's Cassini space probe beamed back from Saturn in 2004. Cassini had picked up the signals from near Saturn's rings, and they sounded slightly musical. But after computer processing the signals sounded to many people like speech. A video of the recording on YouTube has received nearly 4 million hits, and has inspired several others to try it out for themselves.

WORD STATION

pulsar neutron star that emits radiation in regular pulses

What Is a Galaxy?

"Who are we? We find that we live on an insignificant planet of a humdrum star lost in a galaxy in some forgotten corner of a universe in which there are far more galaxies than people."

Carl Sagan, 1980

On a clear night, if you are away from city lights, you can see a great pale band of light stretching across the sky. It's called the Milky Way, and if you look through good binoculars, you'll see it's really countless stars. In fact, the Milky Way is just a sideways view of the great disc of stars our Sun is part of. There are hundreds of billions of stars in the Milky Way Galaxy, yet it is just one of many billions of galaxies in the Universe.

A century ago, astronomers thought the Milky Way was the only galaxy in the Universe. Even with the most powerful telescopes back then other galaxies looked like little more than tiny smudges of light, which astronomers called nebulae. Astronomers argued about whether these were inside the galaxy or outside. The only way to be sure was to find a way to measure the distance to them.

The Andromeda Galaxy is the nearest galaxy to our own Milky Way. Like the Milky Way, it is a **spiral galaxy** and is visible in the sky with the naked eye as a bright fuzzy blob.

The new WFC3 camera onboard the Hubble Space Telescope took this picture of the super-hot Butterfly Nebula.

Finding Andromeda

In 1923, Edwin Hubble (1889-1953) was working with the world's most powerful telescope at California's Mount Wilson. He spotted a Cepheid variable in the nebula known as M31 or Andromeda. Using this Cepheid he discovered the distance to Andromeda and found that it was over 900,000 light-years away (later measurements showed it is actually 2.5 million light-years away). Since the Milky Way was thought to be just 100,000 light-years across, this could only mean Andromeda was far outside it. In fact, it must be an entirely separate galaxy!

Soon astronomers had identified many more galaxies. Although a few nebulae were actually gas clouds in the Milky Way, most turned out to be faraway galaxies, each filled with millions and even billions of stars.

PHOTOS OF THE FURTHEST GALAXIES

In May 2009, astronauts installed a new camera called Wide Field Camera 3 or WFC3 on the Hubble Space Telescope. Like an ordinary camera, WFC3 takes pictures using visible light, but also records snapshots at other wavelengths. By photographing a very small area, the WFC3 camera was able to reveal distant galaxies 13 billion light-years away.

WORD STATION

spiral galaxy flat, disc-shaped galaxy with spiral arms and a large central bulge, such as the Milky Way

These three galaxies make up what is known as the Leo Triplet. The largest of the three, M66, stretches about 100,000 light-years across.

"We don't understand how a single star forms, yet we want to understand how 10 billion stars form."

Professor Carlos Frenk, Director of the Institute of Computational Cosmology, 2004

The big Universe

Astronomers now think there may be 100 billion galaxies in the Universe and have identified them at over 12 billion light-years away. The most distant galaxies so far observed are 13.2 billion light-years away. This means the light from them must have left 13.2 billion years ago. So we are seeing them as they were 13.2 billion years ago, very soon after the Universe was born. It is because astronomers can look back in time like this as they look at ever more distant galaxies that they have gradually pieced together the history of how galaxies formed.

One of the key aims of launching the Hubble Space Telescope in 1990 was to measure distances to galaxies more accurately. The first phase of Hubble's results gave key data from Cepheids to astronomers. From this, they were able to determine very precise measurements of distances to 20 other galaxies. They were also able to determine the size of the Milky Way. This showed that the Milky Way is an ordinary galaxy, very slightly smaller than average.

Cepheids are only clearly visible in nearby galaxies. So to determine how big and how far away other galaxies are, astronomers have to use other distance markers, and all kinds of elaborate ways of measuring distances. Sometimes, they use a special kind of supernova, called Supernova Type 1a. They flash briefly with the brilliance of 15 billion suns, which is why we can see them halfway across the Universe. But they are so short-lived that the chances of finding one in the place where astronomers are looking are small.

The Cerro Tololo Inter-American Observatory is set on top of two mountains in Chile, where the skies are exceptionally clear. Its powerful telescopes play a key role in searching for supernovae.

Astronomer's tricks: How to spot a type 1a supernova

Just after the new moon, when the sky is at its darkest, the Supernova **Cosmology** Project uses a Cerro Tololo telescope in South America to record views of 50 to 100 areas of space containing thousands of very distant galaxies. If any supernovae are visible, they will show up on views taken from a slightly different position three weeks later at Hawaii's Keck observatory. Then astronomers will be ready to follow-up with the Hubble Space Telescope for two months.

The galactic zoo

Most of the brightest galaxies in the Universe are vast spinning discs of stars. The biggest contain 500 billion stars, and all the stars whirl around together as one. Disc-shaped galaxies all seem to have a central bulge, like the yolk in a fried egg. While the "white" is made largely of young, Population I stars, the bulge is mostly older Population II stars.

Some of these disc-shaped galaxies appear flat, with stars spread out evenly in all directions. In most, however, the stars are gathered into spiral arms, which is why these galaxies are called **spiral galaxies**. Barred spiral galaxies have a central bar from which arms seem to trail like water from a spinning garden sprinkler. Our Milky Way is like this. Astronomers think the bar may have formed when the galaxy was disturbed by a close encounter with another galaxy.

What are the oldest galaxies?

Giant **elliptical galaxies** are even bigger than disc galaxies, often containing more than a trillion stars. They can be round, melon-shaped, or anything in between. In some ways, they are like disc galaxies without the disc – just the bulge, full of old Population II stars. But they don't spin as disc galaxies do, although the stars circle around individually like a swarm of bees. They are generally much older, too, and many formed in the earliest days of the Universe.

Known as the Whirlpool Galaxy, M51 is a dramatic example of a spiral galaxy. Both the Whirlpool and its smaller companion can be seen with good binoculars. It is 23 million light-years away.

The Antennae are two galaxies that are colliding as they are drawn together by their mutual gravity. The collision is triggering an enormous burst of young blue stars to form, so they are called starburst galaxies.

The Universe's smallest galaxies are ellipticals, too. These dwarf galaxies contain just a few million stars, and are similar to clusters of stars within larger galaxies, known as **globular clusters**, which contain the oldest stars in the galaxy. Dwarf ellipticals are probably the most numerous galaxies in the Universe. But they are so tiny, we can only see the handful quite near to us.

The final common type of galaxy is an irregular, which has no definite shape. Some may be remnants of other galaxies, torn away when the galaxies collided.

Starburst galaxies

When galaxies get too close, massive bursts of star formation can be triggered, which astronomers call starbursts. In some **starburst galaxies**, stars are forming 100 times faster than in the Milky Way. Many small starburst galaxies look very blue because they are full of hot, young stars, and have hardly any dust clouds. Bigger ones are very red because they are cloaked by huge clouds of dust.

WORD STATION

starburst galaxy galaxy in which a violent event has triggered a burst of star formation

The Milky Way

It's often easy to spot the shape of a distant spiral or elliptical galaxy simply by looking at it through a powerful telescope. But it is much harder with our own Milky Way Galaxy, because we are inside it. Our view is from the side, looking through the flat part of the disc, which means it's like trying to guess the shape of a forest from halfway up a tree in the middle.

The trick has been to map where the Galaxy's stars are, and determine how they move. Using red and blueshift, astronomers have been able to find which stars are moving toward us and which ones away from us, and how fast. By plotting these movements using computers, they have worked out a great deal about the Milky Way.

The Milky Way is the broad band of stars running diagonally across this picture. It is this side on view that we see of the Milky Way from our position on one of the side arms.

Where is the Sun?

We know, for instance, that the Sun lies about two-thirds of the way out from the center of the galaxy. It is on a branch of one of its spiral arms, known as the Orion Arm. Like other stars, the Sun takes about 230 million years to make one complete circuit. The Sun has gone around the Galaxy 20 times since the Sun began 4.5 billion years ago.

From above, the spiral arms of the Milky Way would look very bright because they are full of hot young stars. Spiral arms are hotbeds of star formation, driven by waves of dust and gas that move around the Galaxy at high speeds. As they move through the slower-moving spiral arms, they get squeezed into the clouds from which stars are born. Some astronomers liken these clouds of dust and gas to faster cars on the highway, bunching up into the outer lane as they pass a slower-moving truck, then stretching out as they pass by.

The bent galaxy

For fifty years, astronomers puzzled about why the Milky Way seemed to have a strange kink in the middle. Now astronomer Leo Blitz and his colleagues at Berkeley University in California suggest it is being distorted by the gravity of nearby galaxies called the Magellanic clouds.

This picture shows an artist's idea of what the Milky Way Galaxy and its central bar might look like from far above. The bar is a group of old stars that are moving around together.

Dark Matter and Black Holes

The Universe is a wonderful clockwork of circling objects, with moons orbiting planets, planets orbiting stars, stars orbiting galaxies, and galaxies circling each other. The simplest laws of **physics** tell us that the further out an object is circling, the slower it needs to travel to give it the momentum to keep a steady course.

So astronomers were very surprised to discover in the 1970s that stars at the edges of galaxies are orbiting just as fast as stars near the center. The only good explanation was that the stars are not at the edge of the galaxy as they appear. They are actually held within a much larger disc of matter that we simply cannot see. Astronomers now call this invisible matter **dark matter**.

Astronomers think this image of a distant galaxy cluster shows a ring of dark matter, and provides the best evidence yet that dark matter exists.

This stunning view of the star cluster Omega Centauri in close-up was taken by the Hubble Space Telescope's wide field camera and shows the huge number of ancient stars crammed into the cluster's core.

What shape is the galaxy really?

We now know that all the stars in each galaxy are embedded in huge haloes of dark matter stretching way beyond the visible disc of the galaxy in all directions. In fact, the disc of stars is like a scattering of pepper in between the halves of a gigantic invisible bun of dark matter. We cannot see this bun at all, we just know it is there because of the huge effects of its gravity.

No one knows quite what dark matter is, but it's not just invisible; it hasn't even got enough substance to cluster into stars. It's a bit like an incredibly thin gas that we can walk through without noticing that it's there. The Sun and stars are probably racing through a mist of dark matter all the time. Indeed, dark matter particles may be so tiny that a billion particles may be slipping straight through you right now.

Missing dwarfs

In theory, the Milky Way should be surrounded by hundreds of dwarf galaxies, bound to it by gravity. Yet so far only 20 have been seen. Astronomers have puzzled over these missing galaxies. Now they think these galaxies are actually made of cold dark matter and we simply cannot see them even though they are there. No equipment, however sensitive, can detect dark matter — that's why it's called "dark."

The dark heart of the galaxy

Black holes are places where gravity is so strong that it sucks everything in, including light. They form when a star or part of the galaxy gets so dense it collapses under its own gravity to an infinitely small point called a singularity. Gravity around the singularity is so ferocious that it sucks in space, and even time.

Some black holes form when an old **giant star** collapses in a supernova, for instance. Astronomers can't actually see these black holes. But stars often form pairs or binaries. If one star is a black hole, astronomers might detect the effect of the black hole's gravity on the visible companion star. They may also spot X-rays bursting off matter ripped off the companion star by the power of the black hole.

What is in the center of our galaxy?

Radio telescopes can see through the cloud of dust at the heart of the Milky Way to a region called Sagittarius A* or Sgr A*. In Sgr A*, 20 million stars are packed into a space just 3 light-years across. The stars here are hurtling around so fast that their movement shows up in infrared photographs just a few months apart. Calculations show they must be in the grip of an object 2-3 million times the mass of the Sun. Since this is packed into an area barely twice as big as the Sun, it's clear it must be what astronomers call a "supermassive black hole."

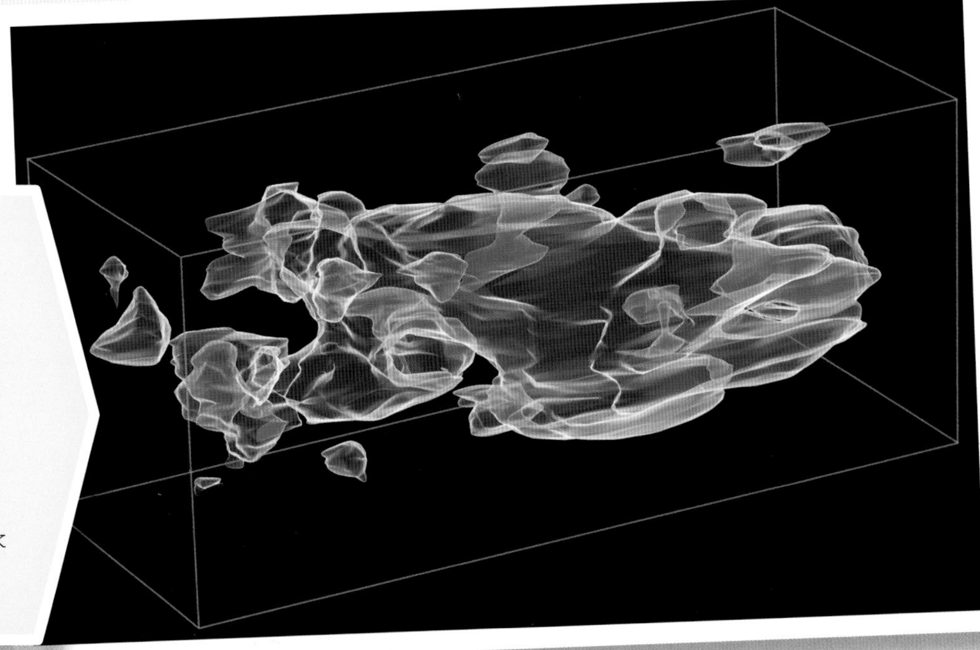

Astronomers used a combination of telescopes to study a portion of the sky and produce this "dark matter map." Studies like this one have helped scientists to better understand how dark matter behaves.

The Chandra telescope's X-ray sensitivity penetrates right to the heart of Sagittarius A* at the center of the Milky Way. It is thought that this incredibly crowded, energetic core is driven by a supermassive black hole.

Astronomers now believe there are supermassive black holes at the center of every spiral galaxy, and it is this powerful black heart that creates their spiral structure and keeps them spinning.

Black holes in the lab

It's impossible to go and study the point at which light disappears in a black hole in space, known as the event horizon. But Ulf Leonhardt and his colleagues at the University of St. Andrews in Scotland managed to create a virtual one in the laboratory using pulses of laser light. That way they don't have to go to space, but can study black holes safely on the ground.

WORD STATION

radio telescope large instrument designed to detect radio waves from space

Galaxies Across the Universe

Stephan's Quintet is a remarkable group of five galaxies (only four are seen here) that seem to be very close together. Indeed they are so close that they seem to be colliding violently with each other, ripping bands of stars off.

After his amazing proof that the Milky Way is not the only galaxy in the Universe, Edwin Hubble made an even more astonishing discovery. He and his assistant Milton Humason calculated the redshift on a number of galaxies. (Remember that redshift is the change in the color of light that indicates a star or galaxy is moving away from us.)

What they found was that the further away a galaxy is, the greater its redshift. In other words, the further away a galaxy is, the faster it is moving away.

In 2008, amateur astronomer Don DeGregori got the chance to look through the powerful Mount Wilson telescope at the planet Mars. At the time, Mars was closer to Earth than it had been for a thousand years.

Why are the galaxies receding?

Astronomers soon began to realize that the Universe is not still, but is expanding continually. The fact that the galaxies are flying away from us doesn't mean that we are at the center and that the galaxies are flying away from just the Earth. No matter where you are in the Universe, they would seem to be flying away in the same way. This is because it is not the galaxies that are moving but the space between all of them that is growing. It's as if they were standing on a 3D rubber sheet that is being stretched in all directions at the same time.

If the Universe is expanding, of course, then there must have been a time when it was very small. If you take today's rate of expansion, you can wind back the clock mathematically. Eventually, you get to a tiny point – the beginning of the Universe. Astronomers talk of the beginning of the Universe, and it is easy to think of this as a mighty explosion which has hurled the galaxies that were born afterwards out in all directions. It has been called the **Big Bang**. But it wasn't really an explosion, and the galaxies aren't actually moving much. It's better to think of space swelling rapidly, like a balloon, carrying the galaxies with it.

ULP Cepheids

Ordinary Cepheids make accurate distance markers only up to 100 million light-years away. Further than that, their light tends to get lost among other stars. In 2009, scientists discovered giant superbright Cepheids called Ultra Long Period (ULP) Cepheids. ULP Cepheids are so bright they can be used to accurately measure distances up to 300 million light-years away.

WORD STATION

Big Bang theory that the Universe started with a tiny point 13.5 billion years ago and has been expanding since

The Hierarchy of Stars and Galaxies

Stars are not alone in the Universe. In fact, the Universe is arranged in a complex pattern of groups. It ranges from individual stars to huge structures containing trillions of groups of stars. This list shows the different levels. Scientists have recently discovered a structure bigger even than the last and biggest group in the list. It's so big they call it the End of Greatness.

This image shows the Rosette Nebula. The central star cluster created a wind that blew a hole through the nebula's center.

1. Binaries and triples: About half of all stars born in the Milky Way are in pairs known as binaries. Most of the remaining stars are in sets of three known as triples.

▼

2. Star clusters: are small groups of stars bound together by gravitational attraction. Stars are born in clusters from nebulae, then slowly drift apart. Clusters vary in size from the 16 stars of the Rosette Nebula to the dense Pleiades cluster of 250 stars.

▼

3. Globular cluster: large clusters of hundreds of thousands or even millions of old stars. Many may have once been dwarf galaxies, incorporated into larger galaxies. These are described on page 39.

▼

4. Galaxies: are very large collections of stars, star clusters, dust and gas nebulae and dark matter. They typically contain between a million and a trillion stars. There are at least 100 billion galaxies in the known universe. They come in three main types: spiral, elliptical, and irregular. These are described on pages 26 and 27.

▼

5. Galaxy groups: are small neighboring groups of galaxies several million light-years apart that move together in space. They might be linked by thin threads of luminous (glowing) gas. The Milky Way galaxy, which contains our solar system, is a member of the Local Group.

▼

6. Galaxy clusters: are huge clusters of galaxies. The average galaxy cluster contains 1,000 trillion times the mass of the Sun and is filled with very hot gas. Within a galaxy cluster, galaxies orbit the center of the cluster just as planets orbit the Sun. The Local Group is part of the Virgo Cluster.

▼

7. Superclusters: Superclusters are gigantic groups of clusters of galaxies, but unlike clusters, they are bound together by gravity. The Local Group is a part of the Virgo Supercluster, with the Virgo Cluster at the center of the supercluster.

▼

8. Walls and voids: The whole Universe is arranged like a gigantic spiderweb. All the stars, galaxies, and clusters are concentrated in vast thin walls or sheets hundreds of millions of light-years across. The biggest is known as the Great Wall, and contains the Virgo Supercluster. In between are gigantic empty spaces or voids.

Active galaxies and quasars

The cores of some galaxies seem to burst with energy. At first, these energetic cores were given various names depending on how they were discovered. Now astronomers think the cores are all the same, so they are given one name, Active Galactic Nuclei (AGNs). AGNs are the brightest objects in the Universe. They get their energy from the supermassive black hole at the heart. As the black hole drags in material, it glows ferociously as it accelerates into a swirling disc around the hole. Sometimes, gigantic jets of glowing **plasma** stream out either side.

What are the brightest galaxies?

The brightest AGNs are known as **quasars**. They can be 10,000 times as bright as the entire Milky Way. Some are so fantastically bright they can be seen blazing 13 billion light-years across the Universe through just a basic astronomical telescope. Some quasars are just 3 billion light-years away, but most are much further. That means we are seeing them early in the life of the Universe. Some astronomers think that spiral galaxies were all quasars before settling down to middle age.

This is an artist's impression of an AGN. The supermassive black hole at the center functions as an AGN's power source.

Globular clusters

When astronomers stare into the distance and see quasars 13 billion light-years away, it is as if they are staring into ancient history. So when they look at distant quasars, they are looking at them when they are very young – even though they may now be 13 billion years old. The oldest stars we can see are quite nearby, in globular clusters, which are clusters of a few million stars within the Milky Way. The amount of the element beryllium in stars slowly rises through time, so astronomers can use beryllium as a kind of clock to reveal the age of a star. In 2004, astronomers found out in this way that stars in the NGC 6397 cluster are 13.4 billion years old.

The M15 cluster is one of the most ancient clusters in the Milky Way. It is thought to be 13.2 billion years old. This means it formed very early in the Universe, just a few hundred thousand years after the Big Bang.

Beginning and Ending

The early Universe was a sea of hot radiation. But after just 200 million years, some astronomers think, the first stars appeared, as atoms clumped together on a giant scale. These early stars were not like most stars today. They are called Population III stars and made entirely of hydrogen and helium. They were so massive that they flared brightly and briefly before blasting themselves apart as supernovae. The shock waves triggered a new generation of stars, the Population II stars. The material that was left over eventually made a third generation, the Population I stars that dominate the skies today.

This image shows an artist's impression of the stages our solar system went through as it formed around our Sun, a Population I star.

These two galaxies have been colliding with each other for about 40 million years. Eventually they will meld into one galaxy.

No one has actually seen Population III stars yet, but they may occur in "faint blue" galaxies — distant blue galaxies formed at the dawn of the Universe. If Population III supernovae did occur, the black holes they left may have merged to form the supermassive blackholes at the heart of spiral galaxies.

How did the galaxies grow?

The merging of black holes may have created galaxies by drawing in dust, gas, and stars to it. Or perhaps the black holes were squashed together by the concentration of dust, gas, and stars. Either way these giant black holes became AGNs that drew in material so ferociously they filled the early Universe with their brilliance.

When all the close material was exhausted, galaxies settled down to a quiet middle age, like the Milky Way today. But they continue to grow. Sometimes, they simply swallow small galaxies. Sometimes they suck in streams of stars from other galaxies that come close. Occasionally large galaxies actually collide. The Milky Way and the Andromeda Galaxy may collide in a few billion years.

Ultra deep space

In January 2004, the Hubble Space Telescope took a picture of a tiny area of the sky that appeared to be completely black to even the most powerful telescope. But by recording for a million seconds, the picture revealed the faint light of distant galaxies. The picture, known as the Hubble Ultra Deep Field, covers an area of the sky no bigger than a grain of sand held at arm's length. Yet it includes over 10,000 previously unseen galaxies. Some were familiar spirals and ellipticals, but others were strange shapes, as yet unexplained.

Edwin Hubble, seen here in 1949. Hubble was able to prove that other galaxies exist outside of the Milky Way. Before this, astronomers believed that the Milky Way made up the entire Universe!

Will it all end in tears?

The scientist Albert Einstein (1879-1955) did some calculations that suggested to him that gravity was so powerful that the Universe must eventually collapse, just like a black hole. But Einstein also believed that the Universe was stable. So he added a figure to his math, which he called the Cosmological Constant. This number balanced out gravity and showed that the Universe was neither collapsing nor expanding. In some ways, the Cosmological Constant was a cheat. Einstein added it in to make his calculations work, not because he had proof of it.

Then, in 1929, Edwin Hubble (1889-1953) made his amazing observation of redshift galaxies that showed the Universe is expanding. This seemed to make the Cosmological Constant irrelevant, since the inward pull of gravity was clearly being outgunned by the outward momentum of the expansion, the momentum of the Big Bang. Einstein said the Cosmological Constant was his big mistake.

Then people began to worry. What will happen when the outward momentum of the Big Bang runs out? To many, there is only one answer. The expansion will go into reverse and the Universe will quickly collapse to nothing in a reverse Big Bang. This is aptly called the **Big Crunch**. Then, in the 1990s, astronomers began to measure the redshift of distant galaxies with very bright supernovae called SN1a. To their astonishment, this showed that the expansion of the Universe is not slowing down but speeding up. So there must be a real force that pushes the Universe outward against gravity with the same effect of Einstein's Cosmological Constant. Astronomers call this force **dark energy**. If there is enough dark energy, the Universe will stretch itself ever further until eventually it tears itself apart entirely. Astronomers call this possibility the **Big Rip**. They estimate that it may happen about 20 billion years from now.

The blob at the end of the Universe

In 2009, astronomer Masami Ouchi spotted a weird object through the Subaru telescope in Japan. It was very, very far away – the fourth most distant object ever seen, at almost 13 billion light-years away. But that wasn't all. No one knows what this shapeless object is, and so they've called it a "blob." It's incredibly bright. Ouchi thinks his blob may be a distant galaxy in the process of feeding on other galaxies, with stars streaming into it.

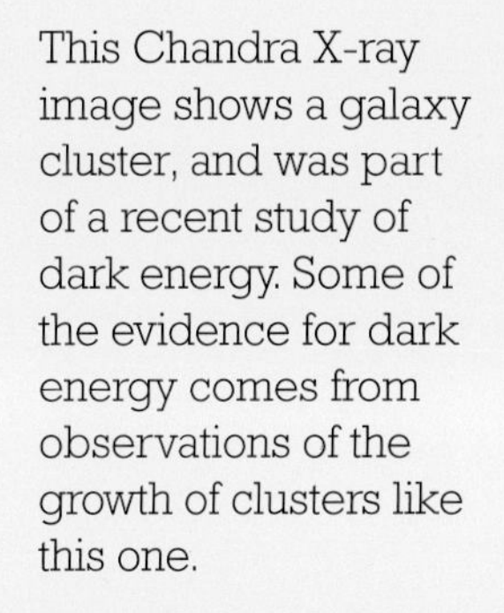

This Chandra X-ray image shows a galaxy cluster, and was part of a recent study of dark energy. Some of the evidence for dark energy comes from observations of the growth of clusters like this one.

Fact File

Star magnitude

The ancient Greek astronomer Hipparchus invented the magnitude system for comparing star brightness around 2,150 years ago. He made the brightest star he could see magnitude 1 and the dimmest 6. Other stars were given scores in between. Today, modern, store-bought binoculars can show stars as faint as magnitude 9. The most powerful astronomical telescopes reveal faint stars of magnitude 30 or fainter. At the other end of the scale, astronomers have spotted very bright stars. But because they are brighter than Hipparchus's brightest star, they actually have negative magnitudes, such as Sirius at -1.46. The Sun is -26.7. A 14 magnitude star is 1 million times fainter than a -1 magnitude star.

The brightest stars	Apparent magnitude	Absolute magnitude	Distance from Earth (light-years)
Sirius	-1.46	1.45	8.6
Canopus	-0.62	-5.62	330
Arcturus	-0.05	-0.3	37
Rigil Kent	-0.01	4.34	4.4
Vega	0.03	0.58	25
Capella	0.08	-0.49	42
Rigel	0.18	-6.81	820
Procyon	0.40	2.68	11
Achernar	0.45	-2.74	140
Betelgeuse	0.45	-5.03	410
Hadar	0.61	-5.35	340
Altair	0.76	2.2	17
Acrux	0.77	-4.23	330
Aldebaran	0.87	-0.64	65
Spica	0.98	-3.62	270
Antares	1.06	-5.45	520
Pollux	1.16	1.09	34
Fomalhaut	1.17	1.74	25
Mimosa	1.25	-3.98	360
Deneb	1.25	-8.75	1,500

Why is the sky dark at night?

If the Universe is filled with shining stars in every direction, as astronomers believe, why is most of the night sky dark? This is called Olbers' paradox. One reason is that some stars are so far away light hasn't had time to reach us since the Universe began. Another is that the expansion of the Universe causes a redshift so extreme that the stars cannot be seen.

Find Out More

Books

Aguilar, David. *Planets, Stars, and Galaxies: A Visual Encyclopedia of Our Universe.* Washington, DC: National Geographic, 2007.

Dyson, Marianne J. *The Space Explorer's Guide to Stars and Galaxies.* New York, NY: Scholastic, 2004.

Ridpath, Ian, ed. *DK Illustrated Encyclopedia of the Universe.* New York, NY: Dorling Kindersley, 2011.

Taschek, Karen. *Death Stars, Weird Galaxies, and a Quasar-Spangled Universe: The Discoveries of the Very Large Array Telescope.* Albuquerque, NM: Univ. of New Mexico Press, 2006.

Websites

www.sciencenewsforkids.org/pages/search.asp?catid=31
A range of stories, including the latest research.

www.space.com/news/
A really good way of keeping up to date with the latest news in astronomy.

www.spacetelescope.org/
The Hubble Space Telescope has brought us some of the most stunning, revealing pictures of the Universe ever seen, and is still doing so. See also the Spitzer Space Telescope.

www.astronomynow.com/
This is the site of the *Astronomy Now* magazine, and has many fascinating articles and stories about stars and galaxies.

http://blog.professorastronomy.com/
A fascinating blog written by U.S. astronomer Kurtis Williams.

Glossary

astronomer person who studies stars in a scientific manner

Big Bang theory that the Universe started with a tiny point about 13.5 billion years ago and has been expanding ever since

Big Crunch theory that the Universe will eventually stop expanding and collapse to nothing again

Big Rip theory that the Universe will go on expanding and eventually tear itself apart

black hole dense, compact object whose gravitational pull is so strong that nothing can escape, not even light. Black holes are thought to result from very massive stars collapsing.

blueshift theory that light from stars and galaxies moving toward us appears slightly bluer as the light waves are compressed

brown dwarf objects that were too small to start the nuclear reactions that would make them glowing stars

Cepheid regular pulsating star

cosmology study of the origin and evolution of the Universe as a whole

dark energy mysterious source of energy introduced to explain why the expansion of the Universe is accelerating

dark matter invisible form of matter that forms most of the mass of a galaxy. Astronomers know it exists because of its gravitational effects.

dark nebula cloud of gas and dust that blocks the light from stars beyond it

electromagnetic spectrum entire range of electromagnetic radiation, ranging from radio waves, to microwaves, to infrared, to visible or optical, to ultraviolet, to X-rays, to gamma rays

element type of matter made up of one particular atom

elliptical galaxy galaxy in which the stars are distributed in an oval shape, ranging from melon-shaped to almost round

fusion joining of the nuclei of atoms of light elements under pressure, releasing huge amounts of energy. The fusion of hydrogen nuclei to make helium is what makes stars shine.

galaxy large assembly of stars, gas, dust, and other matter bound together by gravity

gamma ray very short wave, high energy electromagnetic radiation

giant star star with a radius between 10 and 100 times that of the Sun

globular cluster tightly bound ball of hundreds of thousands, and sometimes millions, of stars about 100 light-years across. They are distributed in the haloes around the Milky Way and other galaxies.

gravity attractive effect that any massive object has on all other massive objects. The greater the mass of the object, the stronger is its gravitational pull.

infrared radiation made of waves slightly longer than red light and just too long to be seen

magnitude scale for comparing the brightness of stars, which can be either its apparent brightness (apparent magnitude) or true brightness (absolute magnitude)

mass measure of the total amount of matter contained within an object

Milky Way our own galaxy to which the Sun belongs. It gets its name because its stars can be seen as a milky band of light across the night sky.

nebula general term used for any "fuzzy" patch on the sky, either light or dark; a cloud of interstellar gas and dust (nebulae is plural)

neutrino tiny subatomic particle that is released during nuclear fusion that moves at close to the speed of light, and interacts very weakly with matter

neutron elementary particle with roughly the same mass as a proton, but electrically neutral. Along with protons, neutrons form the nuclei of atoms.

neutron star superdense core that is all that is left after a supernova typically composed almost entirely of neutrons.

parallax change in the position of an object, or star, in relation to the observer's position or viewpoint

parsec measure of distance in space equal to 3.3 light-years. It is the distance a star must be for its parallax shift as the Earth orbits the Sun to be 1 arc second.

physics study of matter and energy, and the ways in which these interact

plasma state of matter in which all the atoms lose their electrons creating a mixture of free electrons and atomic nuclei

protostar infant star in a gas cloud before it is hot enough for nuclear reactions to start

pulsar neutron star that emits radiation in regular bursts or pulses

quasar incredibly bright, distant objects once thought to be stars but now thought to be gas falling into a supermassive black hole in the center of a galaxy

radio telescope large instrument designed to detect radio waves from space

redshift theory that light from stars and galaxies moving away from us appears slightly redder as the light waves are stretched. The redshift of galaxies that shows they are receding is caused by the expansion of space.

red supergiant huge bright red star

spiral galaxy flat, disc-shaped galaxy with spiral arms and a large central bulge

starburst galaxy galaxy in which a violent event, such as near-collision, has triggered an intense burst of star formation

supergiant extremely luminous, massive star with a radius between 100 and 1,000 times that of the Sun

supernova explosive death of a star

ultraviolet radiation made of waves slightly shorter than violet light and just too short to be seen

variable star star that continually varies in brightness

white dwarf once large star that has exhausted most of its nuclear fuel and collapsed to a very small size

X-ray short wave, high energy invisible radiation

Index